CONCEPTS AND TECHNIQUES IN MODERN GEOGRAPHY No. 7

AN INTRODUCTION TO FACTOR ANALYSIS

by

John Goddard & Andrew Kirby

(University of Newcastle upon Tyne)

CONTENTS

		Page
I	INTRODUCTION	3
II	BASIC CONCEPTS	
	(i) An example of Principal Components Analysis	4
	(ii) An example of Common Factor Analysis	6
III	PRACTICAL & TECHNICAL CONSIDERATIONS	
	(i) The route to a Factor Analysis	12
	(ii) The Data Matrix	13
	(iii) Data Transformations	16
	(iv) The Correlation Matrix	17
IV	THE FACTOR MODEL	
	(i) Principal Components Analysis	18
	(ii) Common Factor Analysis	
	(iii) Communality	
	(iv) The Number of factors	
	(v) The Rotation of Factors	
	(vi) Factor scores	
V	SOME GEOGRAPHICAL EXAMPLES OF FACTOR ANALYSIS	32
VI	CONCLUDING REMARKS	36
	BIBLIOGRAPHY	37

Acknowledgement

We thank R.J. Rummel and W.K.D. Davies for permission to utilise material incorporated in tables 1, 2, 6 and figures 6, 9.

We would like to thank the technical staff of the Newcastle Geography Department for their help in producing a document at great speed: thanks go to John Knipe, Eric Quenet and Sheila Spence. Finally, to Peter Taylor for bludgeoning us into writing this monograph .....

AN INTRODUCTION TO FACTOR ANALYSIS

I INTRODUCTION

It is a basic characteristic of social science that isomorphism, or the existence of simple, one-to-one causal relationships, is rare. An attempt to understand a phenomenon such as, say, "urban growth", thus typically involves an investigation of a series of causally-related variables. Such an examination may be made more rigorous and less time-consuming if our explanatory variables can be considered simultaneously, rather than in a stepwise, "one-after-another" manner. It is this aim that accounts for our interest in multivariate techniques.

There exist a large number of such analyses, each involving slightly different aims and assumptions. Canonical correlation has already been discussed in CATMOG 3 by Clark (1975). Another example is that of multiple regression, where we attempt to "explain" a dependent variable in terms of a series of independent variables; again, an example has been discussed by Taylor (1975). More advanced analyses broaden this framework, and search for inter-relationships among any number of variables. Again, a spectrum of approaches exist, ranging from cluster analysis, which groups variables in ascending order of similarity, (Knox 1974), through to the family of techniques that we shall broadly term factor analysis. Here, our data is simplified, with respect to criteria that we shall discuss below, in such a way that we may create one, or more, new variables, (or factors), each representing a cluster of interrelated variables within the data set.

The reasons why factors are sought have shifted somewhat since the early developments of the technique. Initially, much of the analysis was developed by psychologists, and pioneers such as Burt, Catell, Harman and Thurstone all worked within this discipline. Factor analysis was utilised to relate manifest aspects of mental development, such as introversion, or creativity, to some hypothesised latent dimensions of personality.

A similar type of research design, involving the search for latent dimensions underpinning the relationships between a number of carefully-chosen variables, is to be noted in the pioneer work of Bell (1955) in urban sociology. Here, the hypothesised dimensions of Social Area Analysis were tested, using factor analytic methods applied to seven variables. Three factors emerged from the analysis, which Bell termed 'social status', 'family status' and 'ethnic status'.

Over the last two decades, the advent of the digital computer, and the growing quantification of the Social Sciences, have wrought large changes in the types of user of factor analysis, and consequently the uses to which the technique is put. The speed and storage of contemporary computers have freed multivariate analysis from small data sets and hand-worked solutions of the types outlined by Harman (1960). The change to a growing emphasis upon statistical methods within social science has not always been associated with commensurate developments in theory. Consequently, many applications within subjects like geography involve the use of factor analysis in a hypothesis-creation role, whereby large matrices containing numerous variables are examined for interrelationships.

The simple discussions of hypothesis-testing and creation have been expanded by Rummel (1970) to include ten different uses of factor methods, and these will be illustrated in Section V. In Section II, we shall demonstrate the use of principal components analysis in a data-reduction exercise, and the use of common factor analysis in a hypothesis-testing role. These two techniques are both members of the wider family of factor analytic methods, and although their use is by no means restricted to the applications discussed in Section II, the very different assumptions within the two techniques make them ideally suited to these particular roles. The assumptions themselves will be discussed in detail in Section IV.

II BASIC CONCEPTS

(i) An example of Principal Components Analysis

Having introduced some of the background to factor analysis, let us now consider a small example of the technique in operation.

A readily available illustration is provided by W.K.D. Davies (1971). Here, a small matrix of correlations is presented (table 1), summarising the relationships between five overspill schemes in two towns. The schemes have been compared in a correlation analysis by their scores on fifty-nine attributes, measuring age, population and so on. Any reader unsure as to the meaning of correlation is advised to consult Gregory (1963). The coefficients within the matrix are to be interpreted in the following way. In terms of the fifty-nine attributes, scheme 1, (Aldridge) is very similar to scheme 2, (Aldridge): the correlation is 0.79. In contrast, scheme 3 (Aldridge), has less in common with scheme 5, (Macclesfield); here the correlation is 0.49.

	ALDRIDGE SCHEME			MACCLESFIELD SCHEME		
	1	2	3	4	5	
1	1.00	0.79	0.72	0.75	0.63	Aldridge
2	0.79	1.00	0.56	0.60	0.53	
3	0.72	0.56	1.00	0.50	0.49	
4	0.75	0.60	0.50	1.00	0.75	Macclesfield
5	0.63	0.53	0.49	0.75	1.00	

Table 1 Matrix of correlations: (source Davies 1971) housing schemes in Aldridge and Macclesfield

If our research strategy is to discern some pattern (if any) amongst the five schemes, we can see that we have ten correlations to consider, i.e. 1 with 2, (.79), 1 with 3 (.72) and so on. Even an examination of ten values takes some time, and it is difficult to assess the interrelationships simultaneously. Frequently, a geographical research design involves correlations between dozens of variables, so we can see that the task of discerning general patterns by inspection of the correlation matrix becomes virtually impossible;

(a detailed description of a 57 x 57 matrix is given in Berry & Horton, (1970), pp 323-394).

In this example Davies has performed a principal components analysis upon the correlation matrix in table 1 in order to achieve "an economy of description". The simplest way to describe the strategy is to imagine the existence of a new variable entitled "the degree of similarity between types of overspill schemes". This variable has been calibrated in other studies; consequently we can pick out the elements of similarity within out data. We examine our 59 attributes for scheme 1, and find it closely related to the typical scheme as defined by our measure of similarity: we will call the correlation 0.93. We continue with scheme 2, and give it a correlation of .83, and so on, until we produce the information given in Table 2.

	Aldridge Scheme			Macclesfield Scheme	
	1	2	3	4	5
Component Loadings	0.93	0.83	0.77	0.86	0.81

Table 2 Component loadings: housing schemes in Aldridge and Macclesfield.
(Source: Davies 1971)

Table 2 suggests that all five schemes are closely related to some measure of similarity, and are in consequence all interrelated. If one of our schemes had a value, or loading, of, say, 0.05, we would know that the scheme was not typical, and was unlike the rest of the data set.

In reality, we do not require a variable of similarity. Rather, principal components analysis can examine our matrix of correlation, and pick out the elements of interrelationship as they stand. We can trace the analysis as follows. Firstly, we determine the maximum possible similarity between the five schemes: in this case, if all five were perfectly related (correlation = 1.00) the total would be 25.00 (as there are 25 entries in the correlation matrix). Next, we add up the values in each column of the correlation matrix: thus the total for column one is 3.89. We then sum the five column totals, to give an over all sum of 17.63. This figure summarises the amount of inter-relationship within the matrix; the closer the value is to the maximum possible (25.00 in this case) the greater the degree of similarity between schemes. The column totals show the contribution of individual schemes to the overall sum; by dividing column totals in turn by the square root of the overall value, we can scale this contribution between zero (where a column is composed of zero correlations) and plus or minus one (where a column is composed entirely of perfect positive or negative correlations). It will be remembered that this scale echoes that of the correlation coefficient itself. Returning to our example, we find that the square root of our overall total is 4.200. In the case of scheme one, we divide 3.890 by 4.200, to give us a value of .926, which, it will be noted, compares with the value of .93 given in table 2. We may continue to divide the other four column totals by the overall total, to produce the component loadings already described. By dividing the total of 17.63 by the maximum possible correlation of 25.00, we find that 70.5% of the maximum possible value is achieved. We can use this information in two ways. Firstly, the congruence between the schemes is shown to be over two-thirds of the way towards being total. This tells us, in a precise manner, that the five schemes are highly interrelated. More usefully, we can sum up the entire matrix of 25 entries by the five entries in table 2, whilst losing only 29.5% (100-70.5) of the information within the correlation matrix. We may name the component "degrees of similarity between the five overspill schemes", and note that it picks out schemes 1, 4 and 2 as being most similar to the rest, whilst picking out schemes 3 and 5 as less similar.

Those who have consulted Davies' original paper will have noted that his analysis produces three components, which together account for 92.1% of the variation within the matrix. We can think of the additional components as picking out part of the variations that make up the 29.5% of the explanation not covered by the first component. There are in fact as many components as variables in the analysis, and each one will pick out small patterns of variation that the principal component cannot detect. In section II(ii), we will present an alternative method of considering these subsequent components.

(ii) An example of common factor analysis

As the most widespread use of factor models within geography relate to the 'factorial ecology' mode of analysis, we shall examine in this section the types of census variables seen by the original Social Area analysts as important indicators of certain social dimensions. The data refer to 20 wards of the City of Newcastle upon Tyne, local government units of on average 12 000 people, not the far smaller tracts or enumeration districts typically used in factorial ecology studies. Whilst the major dimensions, such as economic status and family status have been noted in different cities at different times (Robson, 1973), we shall be interested to see how well these patterns of relationship emerge within an industrial city, removed both in time and culture from the original formulations of the social area theory, and studied at a coarse level of detail. An extract from the data matrix is presented in Table 3.

The first three variables approximate to those discussed by Bell (1955) in relation to the concept of economic status. The proportion of unskilled workers is used as a measure of occupational status, the proportion of individuals with A-level qualifications as a measure of educational status, and the mean residential rateable value as an indicator of rent levels. The second three variables relate to family status, and measure fertility (the number of children per 1 000 women of child-bearing age), female activity rates (the proportion of women of working age who are economically active) and housing conditions (the number of shared dwellings).

The first step in our analysis is the choice of factoring design. We have moved away from a simple exercise in data reduction into a more advanced hypothesis - testing situation, i.e. given our six variables, we anticipate finding two factors, one related to variables 1, 2, 3 (economic status) and one related to variables 4, 5, 6 (family status). Because we are investigating a real-world situation, we must take into account matters such as measurement error, and the existence of latent dimensions whose existence we have not hypothesised. We are thus expecting to extract a great deal more information from our data than was the case in the previous example. These differences are reflected in different computational procedures that will be discussed in turn below.

Variable		A Westgate Ward	B Stephenson Ward	C Dene Ward
1.	Number of economically active males in semi and unskilled manual occupations per 1000 economically active males	288	358	95
2.	Number of economically active persons possessing one or more A-levels per 1000 econ. active	31	21	69
3.	Mean rateable value of residential property (£)	93	111	189
4.	Number of children under 5 per 1000 women aged 16-45	538	778	276
5.	Number of economically active women per 1000 women aged 15-60	340	290	348
6.	Number of dwellings lacking exclusive use of amenities per 1000 dwellings	25	99	1

Table 3 Census data for selected wards in the City of Newcastle, 1971.

Due to the different measurement scales of our variables, the data are standardised, by expressing all observations as standard deviations about the respective means. These data are then analysed to produce a correlation matrix of 36 entries.

		1	2	3	4	5	6
Manual Occupations	1	.44	-.52	-.75	.43	-.10	-.10
A Level Education	2	-.52	.81	.65	-.41	.46	.50
Rateable Value	3	-.75	.65	.69	-.59	.29	.09
Young children	4	.43	-.41	-.59	.99	-.70	.34
Econ. active Women	5	-.10	.46	.29	-.70	.35	-.05
Shared Dwellings	6	-.10	.50	.09	.34	-.05	.59

Table 4 Correlations and Communality values, City of Newcastle, 1971

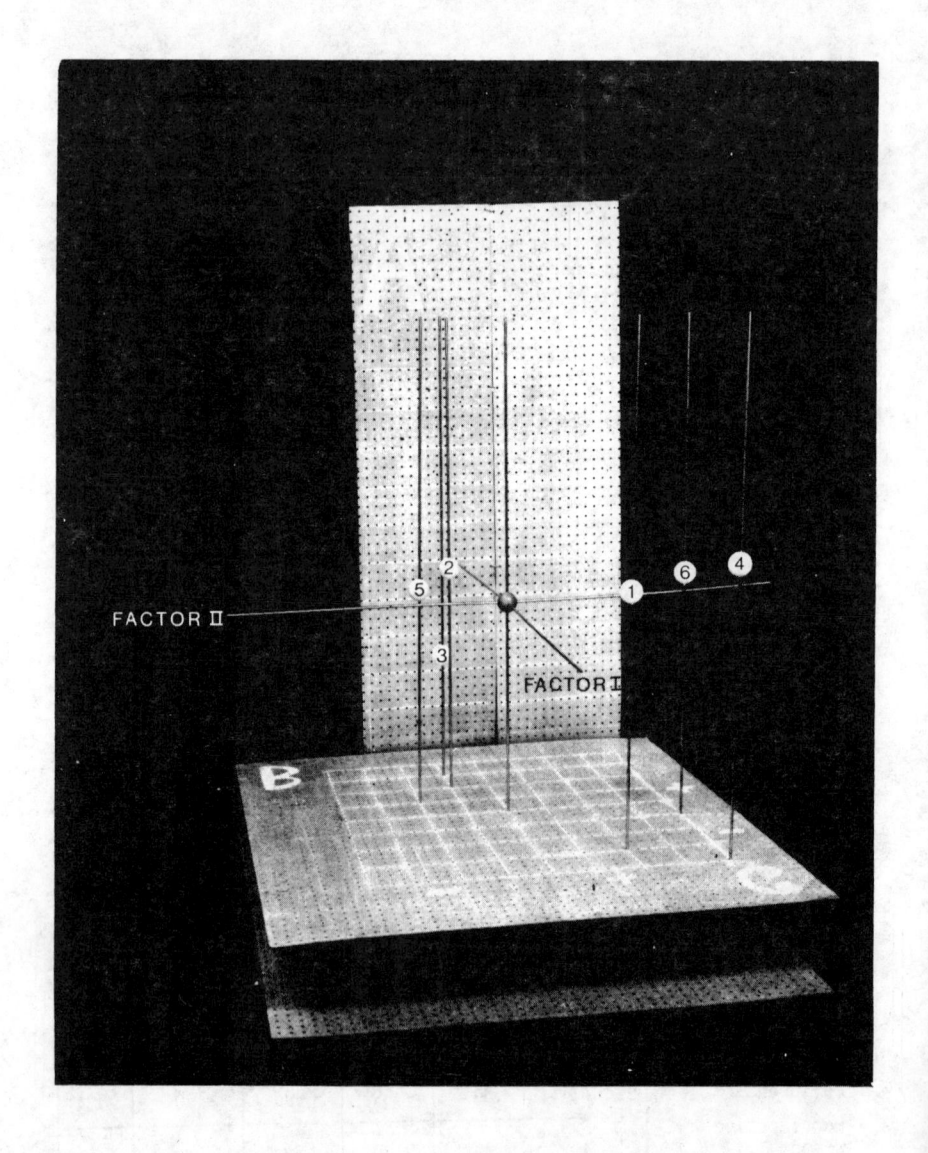

Fig. 1 Social variables in observation space. (Three Wards, City of Newcastle, 1971). (see text for explanation)

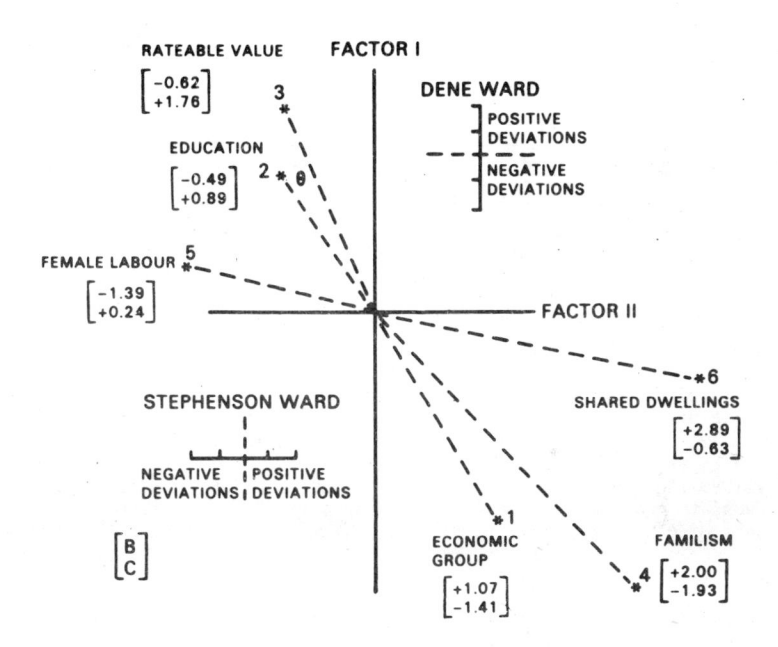

Fig. 2 Representation of data display in Fig. One.
(Scores in standard deviations)

The major difference from the previous example is that the principal diagonal of the matrix, i.e. the correlation of each variable with itself, does not contain values of 1.00. Instead we have inserted estimates of the variation that each variable has in common with the other five variables, and which can be attributable to its correlation with one or more underlying common factors. In factor analysis this is referred to as the communality. The matrix shows that variables 1, 5, 6 have low communalities, and are perhaps accounted for by a dimension, or dimensions, that we are not yet aware of. The methods used to determine the values of the communalities are discussed in Section IV.

The simplest way to understand the factoring procedure is to display the variables and the resulting factors in a three-dimensional format (Figure 1). Here, the variables are located in object-space, i.e. each of the axes A, B, C represents one of our twenty wards; the location of the variables is with respect to their value about the mean, represented by the darker, central sphere. Further details of the physical model are provided in Kirby (1976).

Despite the small sample of wards, we can immediately see that a pattern of relationships exists. To the left of the model, we can see a cluster, formed by variables 2, 3, 5 (education, rateable value and working women) whilst to the right we find variables 1, 4, 6 (occupation, fertility and shared dwellings). We may continue to define this pattern in a precise way, by examining the geometry of the interrelationships. Firstly, we may imagine a series of lines, usually termed vectors in this context, radiating out from the origin to the variables, like the spokes of a wheel about the hub. The closer the spokes, the more closely related are the respective variables. We can quantify this by measuring the acute angle between the 'spokes', and converting this to its cosine value. The cosine and the product moment correlation coefficient that we have already used are interchangeable, and in this way we can determine the strength of relationship between our variables. For example, let us measure the acute angle Θ between the vectors connecting variable 2, and variable 3, (Figure 2). This gives us a value of approximately 10°: the cosine of this angle is in excess of 0.9. Consequently, we may now state that the relationship between variables 2 and 3 is a correlation in excess of 0.9, i.e. strongly positive. If we return to table 3, we find that the correlation value measured over all 20 wards is 0.65, again indicating close relationship.

Rather than continue to measure a whole series of acute angles, between all the pairs of variables, we may summarise these inter-relationships with the use of two further 'spokes', representing the resultants of the data vectors. Child (1970) suggests a useful analogy, whereby we may imagine the resultant of a series of vectors to be a generalisation of the divergent directions in much the same way that the handle of a half-opened umbrella represents a middle way, or 'average' direction. These resultants are the factors that we are seeking, concisely stating the pattern of relationships within the vectors in the same way that we saw components summarising a correlation matrix.

Figures 1 and 2 show the two factors in place. The factor running north-south is the principal factor. It can be seen that if the angles between the factors and the original test vectors are measured as before, we can produce the correlation values, or loadings, between the variables and the factors. Thus, variables 2, 3 load most strongly on the first factor, as does variable

1, but with, of course, a different sign. Variables 4, 5, 6 are less highly correlated.

The second factor normally displays less interesting patterns, in so far as the location is primarily determined by the first factor. It is typical to find the second factor cutting across a swarm of data, producing a so-called bi-polar factor, where loadings are equally matched, positive-negative. In this example, variables 5, 6 both load quite strongly on this factor.

In a realistic research situation, we would not attempt to produce meaningful factors using only three cases, and for this reason we have not included in this section details of the loadings discussed above. Nevertheless, the general patterns found using these three cases are representative of the results of the analysis using all the twenty wards. These are displayed in Table 5, in conjunction with detailed information on the interpretation of factor results.

	VARIABLES	FACTORS Loadings		COMMUNALITY Sum of Squared Row Loadings
	EXPLANATION	Loadings: Correlation between Variable and Factor. Squared Loadings: Prop. of variance of each variable accounted for by factor.		Communality: Proportion of variation accounted for by both factors I & II.
		I	II	
1	Manual occupations	-.65	-.14	.44
2	A Level Education	.79	.42	.80
3	Rateable Value	.82	.10	.68
4	Young children	-.81	.57	.98
5	Econ. Active Women	.53	-.25	.34
6	Shared Dwellings	.11	.76	.59
	% Variation Accounted for	44.5%	19.5%	63.8%
	Eigenvalues: Sum of Squared Column Loadings	2.67	1.17	Total variation Accounted

Table 5 Output from factor analysis, City of Newcastle, 1971

As we noted above, the first factor is defined by positive association with the variables education, rent, and female labour and negatively by variables indicating low social class and familism. The second factor is characterised by low loadings, except in the case of shared dwellings. An interpretation of these results suggests that whilst two dimensions of economic status and family status were hypothesised, we have here produced a general dimension of social differentiation, picking out low social class and large families on the one hand, and educational attainment, housing status and economically active women on the other. The variable measuring housing quality appears to be unconnected with these other data items, loading as it does highly on the second factor. We may note that the existence of economically active women has become an indicator of higher social status, as opposed to lower social status as hypothesised. Our interpretation of the second factor, which picks out shared dwellings, must rest to some degree upon our knowledge of the city, for it seems likely that the scale of the analysis has failed to pick out the many pockets of shared dwellings that exist in Newcastle, and which are closely related to the distribution of low economic status groups and variations in rateable values. Such problems of interpreting, or labelling factors, are very common.

To sum up this analysis, we must reject our hypothesis that two dimensions of urban differentiation exist. Whether this is a function of our data base (i.e. Newcastle) or the scale of the analysis is a point that can only be determined by further analysis of other cities and of Newcastle, using more detailed data for enumeration districts. Attention must also be paid to the fact that our level of explanation is quite low: the two factors considered account for only 65% of the variation within the correlation matrix. Once more, we must suggest that further research is necessary to pin down whether this rather poor rate of explanation is due to the distortions of the scale of the analysis, or the existence of latent dimensions of differentiation that we have not hypothesised.

III PRACTICAL & TECHNICAL CONSIDERATIONS

(i) The route to a factor analysis

The two examples we have given suggest that in searching for order through factor analysis we can attempt two separate although related tasks. First, we can eliminate the redundancies in the original set of data by attempting to summarise the variation in this set in terms of a smaller set of variables (or factors) which are a combination of the original variables. We then attempt to identify or name the underlying dimensions in the data set by examining the way in which the original variables have been combined in the process of making this summary. The way we combine a set of characteristics such as occupation, education and income measured over a set of census tracts, may lead us to a basic underlying dimension that we might name "social status". Each of the original census tracts may be scaled along this dimension by computing its score on the basic factor. In more everyday language, we can combine many of the detailed characteristics of a number of motor cars to arrive at some underlying notion of "performance" and hence derive a multivariate scale on which each individual car can be rated.

The second way in which we search for order amongst a set of phenomena is to postulate a limited number of variables or factors which are responsible for producing the inter-relations between the original variables. For instance, we could hypothesize that a force called "economic development" generates typical levels of G.N.P.,urban growth rates, levels of educational attainment, etc. and test for the existence of such a basic dimension amongst a large set of variables measuring these characteristics.

In the first instance we have used factor analysis within an inductive framework and in the second we have used it deductively. Usually factor analysis is regarded as a classical inductive method. Indeed, in most instances it is used as an exploratory tool for unearthing the basic empirical regularities in a set of data. Used in this way, the technique is a very powerful descriptive device.

Methodological questions of induction and deduction are obviously not independent of the problem under investigation. All of these matters closely impinge on the question of factor analysis research design. This is because factor analysis embodies a number of alternative procedures; indeed the term "factor analysis" should be regarded as a generic name for a family of multivariate techniques. Any one problem might involve several of these procedures. Consequently the researchers will need to specify an appropriate factoring design. Figure 3 represents a flow diagram of a possible factor analysis research design. In many instances not all of the steps will necessarily be executed. The analysis may terminate at an early stage or certain steps may be by-passed. Also, different methods of common factor analysis may be adopted which do not involve all of the procedures suggested in Figure 3, while other approaches provide statistical tests for the number of significant factors. Although the diagram is not a completely general model of factor analysis research design, it does cover the two most frequently used approaches in geography; it also provides a useful framework for the ensuring presentation of issues.

Before proceeding one obvious point should be stressed: the variety of alternative choices available at different stages in completing a factor analysis means that each choice, or combination of choices, will produce a different end product. Although there are a number of guidelines as to the most appropriate choice at each stage, an element of subjectivity will inevitably be involved. In many instances there may not be a single "right" answer. As the ultimate test of the success of the factor analaysis may be the interpretability of the final results, producing an adequate factor analysis is as much an art as a science. However, one important scientific criterion can be satisfied and that is reproducibility of the result by another investigator - provided that is, every choice that has been made is carefully documented in writing up the results.

(ii) The Data Matrix

The first choice facing us concerns the selection of variables. The output of the factor analysis will obviously depend on the nature of the input. This is true of every statistical analysis, but because of factor analysis's ability to handle large numbers of variables there is a danger of this point being overlooked. We may be tempted to include all of the variables available, for example in the census, without carefully considering the mix of underlying dimensions that these variables cover. So it is not surprising that

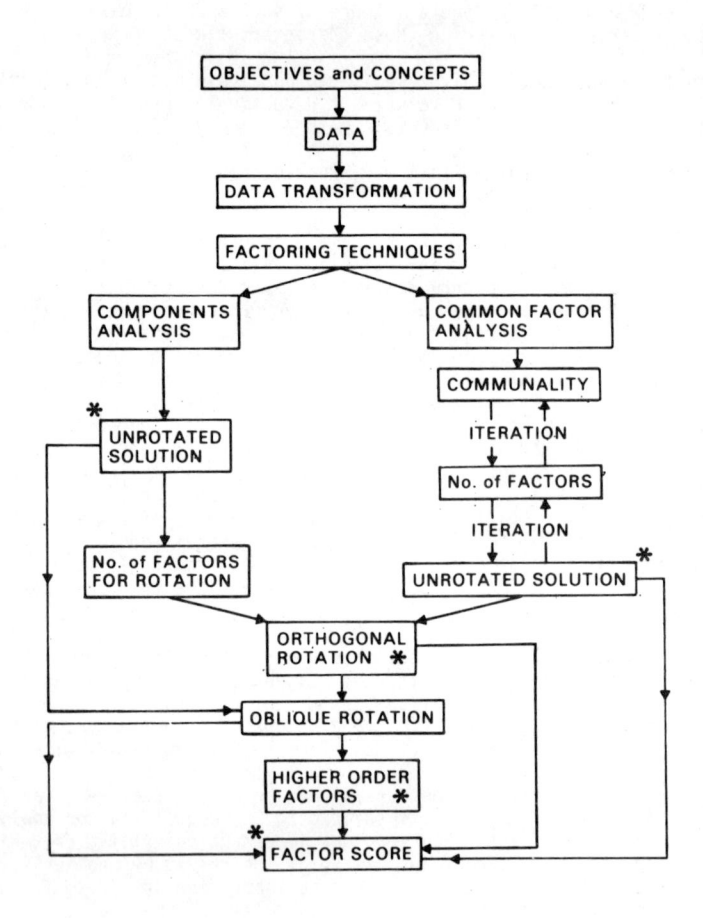

Fig. 3 Routes to Factor Analysis. (*Possible termination points)

factor analyses of British cities often extract housing conditions as a
principal factor underlying spatial variations in their internal structure,
while in the American context an economic status dimension is usually more
important. This may simply reflect the fact that the British census contains
more indicators of housing conditions while the American census has more
socio-economic indicators. Having selected our variables we may be faced with
the question of the scale at which to carry out the analysis, that is, the
selection of observations. For example, in a study of urban structure, should
we analyse the data at the level of the enumeration district (with an average
population of 500) or wards (with an average population of 12 000)? The results may differ significantly for these two scales.

The next step involves arraying the data in matrix form. Typically this
would consist of n cases (census tracts, industrial plants, cities, etc.)
over which m variables or characteristics are measured (Figure 4). These variables may be measured in any units (dollars, people, tons, etc.) and scaled
in any way (interval, ratio, nominal). It should be stressed that wherever
possible the number of cases (n) should exceed the number of variables (m).
Re-examination of Figure 1 should immediately suggest the reason - namely that
the observational space set up to contain the variables would have fewer
dimensions than the number of variables. The position of any vector in this
space could not therefore be uniquely determined. Such a matrix might be referred to as a <u>spatial structure matrix</u> and form the initial data for a multi-
factor uniform regionalisation (i.e. a grouping of the n cases on the basis
of similarity in terms of less than m structural dimensions). Examples of
such matrices are provided by King's study of Canadian urban dimensions,
where the observations consisted of 106 cities and the variables 50 socio-
economic characteristics of these cities (King, 1966), and Goddard's study
of central London where the observations consisted of 260 city blocks and the
variables 80 different types of office employment (Goddard, 1968).

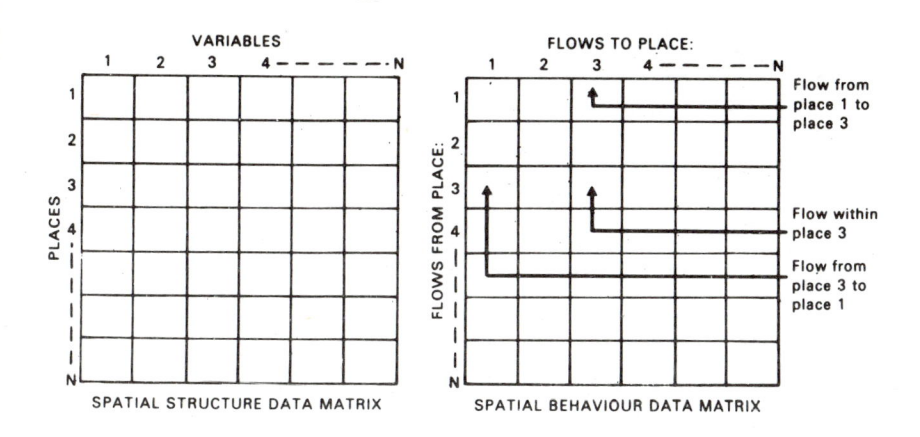

Fig. 4   Spatial structure and behaviour matrices

Alternatively, the new data matrix might consist of an n x m transaction matrix in which the entries represent the flows between places (Fig. 4). Such a matrix need not be symmetric (i.e. the flows from A to B may be greater than the flows from B to A). An example is given by Goddard's study of taxi flows in Central London (1970). Such matrices might be referred to as <u>spatial behaviour matrices</u> and form the initial data for a functional regionalisation (i.e. a grouping of places on the basis of functional ties).

Let us consider the structure matrix. Essentially two questions can be asked of the variation expressed in this matrix;

1. Which variables possess common patterns of variation over the set of observations? This is called R mode analysis (column-wise comparisons)
2. Which cases possess similar profiles of scores over the set of variables? This is called Q mode analysis (row wise comparisons).

Whilst geographical literature has tended to concentrate upon the relationships between variables, it will be remembered that our introductory example was of relationships between observations: as Davies observes "one would expect geographers to be more directly concerned with areal differences and similarities" (Davies 1972, p. 115) and hence Q mode analysis. However, in many geographical studies the number of observations (areas) frequently exceeds the number of variables, thereby precluding Q mode analysis.

If the data matrix had been a cube with different time periods constituting the third dimension, numerous other slices of the matrix could be taken for analysis - for instance, where dates are the observations.

In the case of the transaction matrix, R mode analysis would compare destinations in terms of similarities in the origin of the flows and Q mode analysis would compare origins in terms of similarities in the destinations of their flows.

(iii) Data Transformation

Having arrayed our raw data in matrix form, some transformations may be required to take account of particular peculiarities of the data, such as different measurement scales, or variations in the size of areal units. For example, it is common to express raw variables as ratios of some base population in an attempt to eliminate size effects. Thus to standardize for variations between census tracts in the number of households we might express each variable as a ratio of the total number of households in the area (e.g. the proportion of households sharing a dwelling). If this were not done, the largest amount of variation in the data could arise simply from differences in the size of areal units, and this variation would be reflected in the first factor. However, in making such transformations we should be aware of two dangers. First, of false inferences that could be drawn from interpreting the results in terms of the original and not the transformed variables, and second, of the danger of <u>spurious correlation</u> resulting from dividing a number of variables by a single common denominator. For example, the correlation between the proportion of households renting their accommodation and the proportion owning their accommodation may be more strongly negative than that between the absolute number of households in each category.

Some transformation of the data matrix may also be required by the factor analysis method that is to be employed at a later stage. Data transformation may be divided into those that are applied to only single variables or subsets of variables and those that are applied uniformly across the data matrix.

Distributional Transformations: It has commonly been assumed that factor analysis requires the underlying distribution of the data to be of a multivariate normal form; this implies that not only each variable is normally distributed but that the relationship between all pairs of variables is linear. Some workers argue that normality is essential only if questions of statistical inference are involved. While this is true it should be stressed that a sufficient condition for the correlation coefficient (on which subsequent factoring is based) to be a true measure of the statistical association between two variables is that the bivariate distribution of the variables be normal. Therefore if product moment correlation coefficients are used as a measure of spatial association between the m variables it is advisable that these variables be transformed such that the relationship between all pairs is linear. Transformation to normality for each variable will increase the chances of near linear relationships, but will not guarantee this.

Matrix Transformations: If the data are measured in a large variety of different units - for example some in terms of dollars, some in terms of population, some in percentages, it is necessary to standardise the data into similar units so that meaningful comparisons between the distributions can be made. The most frequently used standardisation involves expressing the original observations in standard score units. The standardisation transformation subtracts the mean of the data for a variable from the original data and then divides by the standard deviation. The effect of the transformation is to remove the difference in mean and standard deviation between variables from their co-variance; each variable now has a mean of zero and a variance of one. (Standardisation creates some problems as it generally gives equal weight to each variable in the analysis, although some variables may be far more significant - in terms of exhibiting a wider range of variations over the set of individuals - than the others). The formula is:

$$Zij = \frac{(Xij - \bar{X}j)}{Sj}$$

Zij = standard score of i th observation on j th variable

Xij = raw data of i th observations on j th variable

Xj = mean of j th variable

Sj = standard deviation of j th variable

(iv) The Correlation Matrix

Such a standardisation is effected when the product moment correlations are computed between each of the variables in the data matrix. In the case of a spatial structure data matrix the distributions of each pair of m variables is compared over all of the observations to produce a symmetrical m x m correlation matrix. The correlation coefficients indicate similarities in the profiles of each variable observed over the set of areal units. In the case of the transaction matrix (R analysis), correlations between columns compare areas in terms of similarity in the profiles of their trip origins. No account is taken of similarity in magnitude. In the analysis of transaction matrices it might be important to compare variables in terms of similarities in the volume of, say, trade received. In this case, an alternative to the product moment correlation coefficient, the pattern magnitude coefficient might be employed; this takes account of both sources of variation. If, however, the flow matrix is symmetrical (i.e. flows from A to B equal those from B to A) raw data for each variable may be scaled to range from 0 to 1 and analysed as a correlation matrix (Rummel, 1970).

The type of coefficient employed in the transformation to a correlation matrix will also depend upon the way in which the data is scaled. In the case of nominal data the phi coefficient may be employed: a recent example of the analysis of binary data is provided by Tinkler (1972,1975) & Hay (1975). In the case of ordinal data some rank order coefficient and in the case of scale data (i.e. data with a restricted range) some non-parametric coefficent should be applied. Any of the matrices of coefficents may be factored. Alternatively, and this is particularly relevant in the case of transaction matrices, the raw data may be directly factored without computing the correlation coefficients. (Horst, 1965, Russett, 1967).

Much useful information can be gained from careful examination of the correlation matrix. Identification of clusters of highly inter-correlated variables can sometimes provide a guide as to the number and character of the underlying common factors. All too often in geographical studies no attention is given to the correlation matrix.

IV THE FACTOR MODEL

In the introductory sections we touched upon the differences that exist between principal components analysis and common factor analysis. The two examples discussed have provided an example of each model. We shall now contrast the two forms of analysis in greater depth.

(i) Principal Components Analysis

Principal components analysis is a data transformation procedure applied to the raw data matrix Z. Essentially it seeks to replace the m columns of Z with new columns, which, if represented geometrically as vectors, would be mutually orthogonal. We seek to define a new set of variables or components that are uncorrelated and where the definition of these components is in terms of m coefficients relating them to the original variables.

A geometric interpretation for two variables is possible. Consider a conventional scatter diagram in which the co-ordinate axes represent the two variables Z_1 and Z_2 and the location of the observations are defined by their scores on each of these variables (Fig. 5). Notice we have defined a variable space and not an observational space as in Fig. 1. The oval shape of the swarm of points is indicative of the correlation between the two variables. A perfect circle would indicate zero correlation; in fact we could rotate the variable axes to such a position that the points were in a circular swarm and the angle between these axes would again indicate the degree of correlation. The two representations can therefore be transformed into one another.

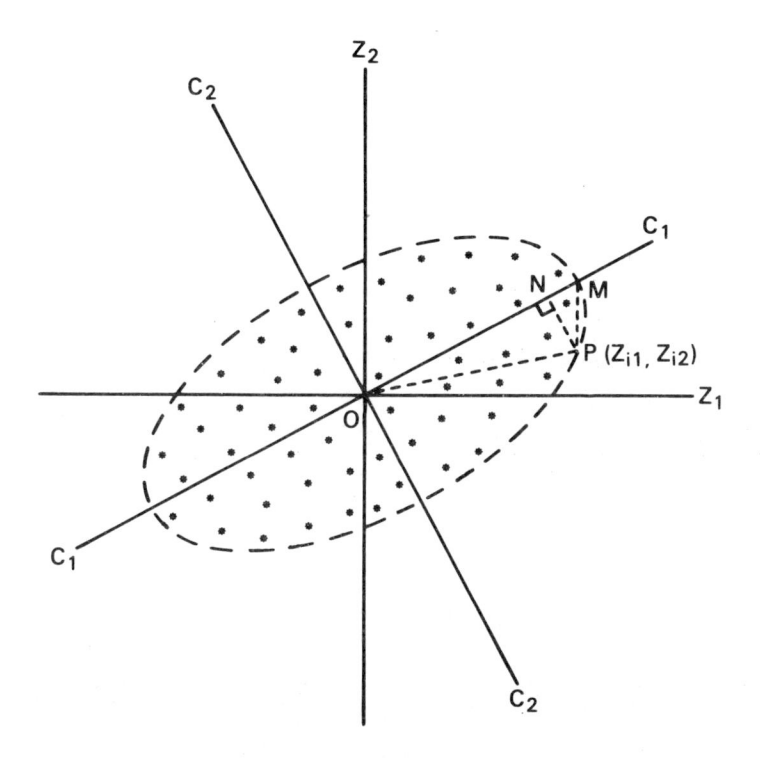

Fig. 5 Components as the major and minor axes of an elipse

We can draw a line C_1-C_1 through the scatter of points such that the sum of the squares of the perpendicular distances PN are minimised (in regression, it is the sum of the squares of the distances PM which are minimised); C_1-C_1 is thus the line which accounts for the greatest amount of common variation in the two variables. It is the axis representation of the first component. Unless all points lie on a perfectly straight line, C_1-C_1 will not account for all the variance in Z_1 and Z_2. However, if we draw a line C_2-C_2 orthogonal to C_1-C_1, the two new lines now represent all the variance in the two variables, since we may define new co-ordinates for the i th point with respect to the new axis. Thus:

$$C_{i_1}^2 + C_{i_2}^2 = Z_{i_1}^2 + Z_{i_2}^2 = OP^2$$

C_2-C_2 is the axis representation of the second component.

The above procedure can be readily extended to more than the two dimensions.
The following properties hold:-

1. The first component always accounts for the greatest amount of variance in the original scatter of points.

2. When the number of variables is greater than two then the second component accounts for the greatest proportion of residual variance (i.e. that remaining after that due to the first component is removed).

3. There are exactly the same number of components as original variables and the components are ordered in terms of the proportion of variation in the original data set accounted for by each component.

4. In cases in which the number of variables is large the great majority of the variance in Z is accounted for by a relatively small number of components. This achieves a parsimonious description of the data.

The principal axes of a data matrix can be conveniently solved for algebraically. The question is what transformation will rotate the variable axes so that they lie co-linear with the principle axis. These transformations are provided for by solving for the eigen-values and the associated eigen-vectors of the full correlation matrix (i.e. with "ones" in the diagonal; 1.0 equals the variance of each standardised variable. Thus the total amount of variance to be accounted for is equal to the sum of the diagonal elements of the correlation matrix). The eigen-value indicates the length of each of the axes, i.e. the amount of common variance accounted for by each axis. Gould has provided a useful geographical interpretation of eigen-values (Gould, 1967). Dividing the eigen-value by the number of variables indicates the proportion of the total variance in the standardised data matrix accounted for by each component. By scaling each eigen-vector by its associated eigen-value we reduce the length of the eigen-vector proportional to the length of the principal axis it measures. The elements of the eigen-vectors contain the transformation coefficients for the variables that rotates each variable axis by the angles α and β so that it lies along the principal axis (Fig. 6).

These coefficients are the correlations or loadings of each variable on the successive principal axes. With m variables there will be m eigen-values and associated vectors. The first will account for the largest amount of variance in the standardised data matrix, the second the largest amount of variance in the residual matrix after the first eigen-value has been extracted, and so on. Together the m values will account for the total amounts of variance in the original data. It is thus possible to write a series of equations, one for each variable, to completely specify or recreate the original variables in terms of the new components:-

$$Z_i = A_{i1}F_1 + A_{i2}F_2 + \ldots\ldots\ldots A_{ip}F_p$$

where:

i = 1, 2, ... m and p = m and p is number of components; m is number of variables.

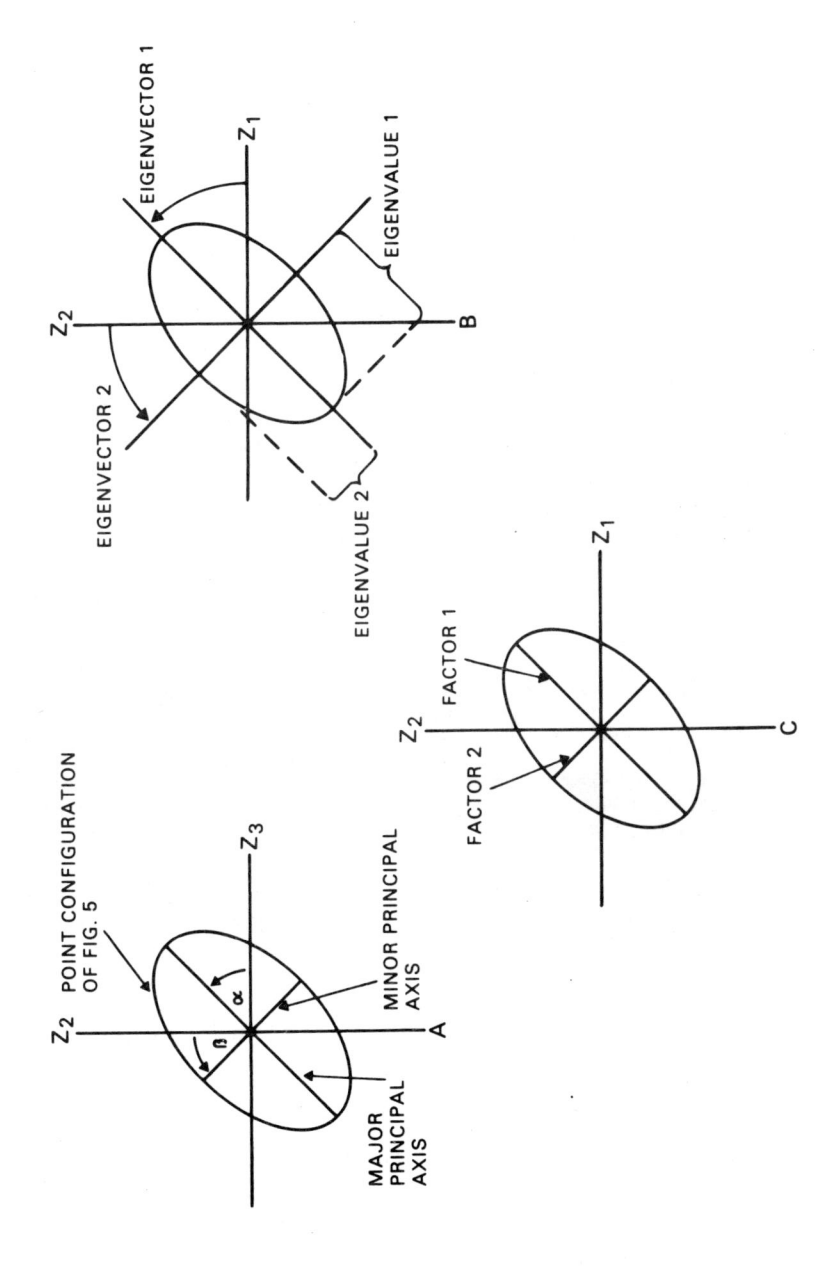

Fig. 6 Geometric interpretation of eigenvectors and eigenvalues. (Source: Rummel, 1970)

Examinations of the coefficients in these equations (the component loadings or the correlations between the component and the variable) helps in the naming of the component.

Components analysis alone is most useful when there is a major dimension of variation in a data set which itself identifies a clearly discernible concept and within which there are no important sub-concepts. Berry, for instance, has used each country's score in a cross national components analysis of socio-economic variables to place it on a multi-variate scale of economic development (Berry, 1960). Similarly, Gould in several studies has used components analysis to extract the group image in studies of residential preferences. (For example, Gould and White, 1968). From an n place x m individual matrix of preference scores, he extracts the principal components which usually accounts for a substantial proportion of the original variance. Mapping of components scores for the n places reveals the group's common preference surface.

A further important use of principal components analysis is as a straightforward data transformation to meet the assumptions of other techniques. In particular multiple regression assumes that the predictor-variables, the so-called independent variables, are uncorrelated. If the independent variables display strong inter-correlations then components analysis can be applied to orthogonalise the data; if sufficient variance is explained the scores of the observations on the leading component can be entered into a simple regression in lieu of the original intercorrelated variables. If interpretation is not possible or if explained variance is low, the scores from more than one component can be entered in a multiple regression. An example is provided by Britton's gravity model study of the freight hinterland of Bristol where several mass factors were highly inter-correlated and had to be transformed in a principal components analysis before calibrating the model through multiple regression. (Britton, 1968).

It should be stressed that principal components analysis is essentially variance orientated. If we return to the geometric interpretation, a configuration of m variables will describe a hyperelipsoid in m dimensions (in three dimensions a rugby football provides a good analogy of the shape). The first component will be inserted along the principal axis of variation regardless of the existence of distinctive clusters of variables in this space. All variables included in the analysis, no matter how randomly they are related to the rest of the data set, would influence this principal axis of variation. Most variables will therefore load fairly high on this component. In many geographical studies, when the data has not been weighted for differences in the sizes of areal units, this first component can often be identified as a size factor. The second component will be inserted orthogonally to the first and will often be bipolar, i.e., with both high negative and positive loadings, which creates difficulties in interpretation. (i.e. in the distinguishing of independent clusters of variables)

If interpretation of the underlying dimensions is sought, rather than simple parsimonious description of the data, two other procedures present themselves. First, rotation of the principal axes to some more meaningful position that better describes the initial clusters of variables, and secondly, the adoption of a fundamentally different factoring technique, namely common factor analysis, which concentrates on the major patterns of variation and chooses to ignore the remainder.

r=0.6

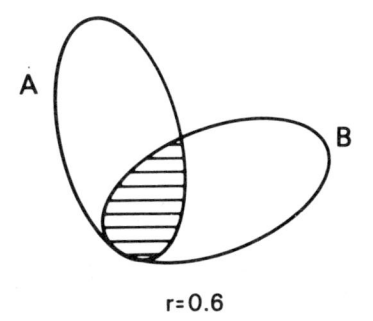

Fig. 7 Elements of common variance in bivariate correlation

(ii) <u>Common Factor Analysis</u>

Common factor analysis begins with a different set of assumptions from principal components analysis. It is assumed that the variation in a given variable is produced by a small number of underlying factors and by a variation unique to itself. Consider the correlation between two variables: this can be interpreted in terms of elements of common variance. In psychology those areas of overlap are considered as brought by basic underlying influences, i.e. the correlation (r) between A and B is due to the correlation between A and an underlying factor and B and the same underlying factor (Fig. 7). Viz:-

$$rAB = rAF \times rBF$$

The proportion of the variance in a variable accounted for by the common factor is called the <u>communality</u> and the remainder of the variance its <u>uniqueness</u>. It is assumed that all the co-variation between the variables is produced by the underlying factors, the unique components of the variables being uncorrelated:

$$Z_i = A_{i_1}F_1 + A_{i_2}F_2 + .. A_{ip}F_p + A_iU_i$$

where: $i = 1, 2, \ldots m$ $p < m$ and p is number of factors:
U is error term, one for each variable.

In other words we have one equation of the above form for each of the variables, the F1, F2 to Fp are the common factors and the Ui is the unique component of the i th variable. The communality of the i th variable hi is the variance in Zi due to the common factors:

$$hi^2 = \sum_{k=1}^{p} Aik^2$$

The uniqueness of the i th variable is equal to 1 minus its communality. Provided that the factors are orthogonal, the coefficients Aik still represent the correlation between the i th variable and the k th factor.

The use of this model poses two problems. First one does not initially know the magnitude of the unique components of variation for each variable and hence its communality. And second, one does not know the number of common factors which are appropriate. In psychology the number of basic dimensions can often be hypothesised from theory but in most goegraphic applications this is rarely the case.

(iii) <u>Communality</u>

Standard procedures have now been developed for estimating initial communalities and comparing these with final communalities and reiterating the process until there is no significant change. The lower bound for this estimate has been shown to be provided by the square of multiple correlation resulting from the regression of the m th variable on all the other m - 1 variables. The communality estimates are inserted in the diagonal of the correlation matrix. We are thus assuming that the variance of a variable to be explained by the common factors is less than 100%. The analysis is pursued in the same way as that outlined for principal components analysis. Final communalities (i.e. the actual amount of variance explained by each of the components) are reinserted in the diagonal of the correlation matrix and the processes reiterated until some satisfactory convergence is achieved.

(iv) <u>The number of factors</u>

Using the principal axes method, factors are extracted successively - the first accounting for the maximum amount of variation; the second axis is then extracted from the residual correlation matrix and so on. The question then arises as to when to stop extracting eigen-vectors. Several rules of thumb have been suggested as to the number of factors, but the ultimate test is the interpretability of the resulting factors. One may initially opt for a satisfactory amount of explained variance and this is usually provided by all those eigen-vectors with eigen-values greater than 1.0 - i.e. only those factors that account for more than their proportionate share of the original variance. Within these limits a rough guide to interpretability is provided by plotting factors against eigen-values and seeking a distinct break of slope (Fig. 8). Figure 8 provides one such "scree diagram" from a study which sought to identify the number of distinct spatial clusters of office activities in the City of London (Goddard 1968). It will be seen that beyond the sixth component each additional component accounts for approximately similar and trivial amounts of variance.

(v) <u>The Rotation of Factors</u>

In some investigations we may only be concerned with obtaining a parsimonious description of our data while in other instances we may aim to identify distinctive clusters of variables. If the latter is our aim it may be necessary to carry out a rotation of the original factors; this is because the initial solution will identify the principal patterns of variation and not necessarily distinctive clusters of variables.

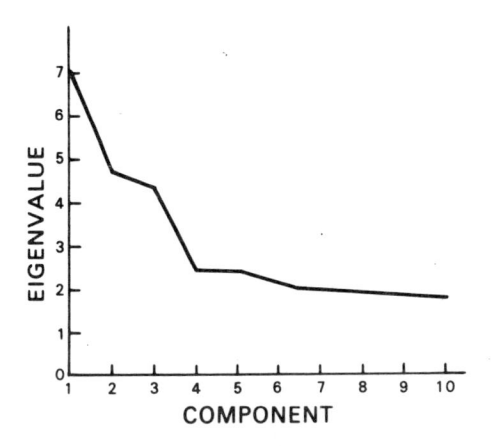

Fig. 8 Trivial and non-trivial eigenvalues. (Source: Goddard, 1968)

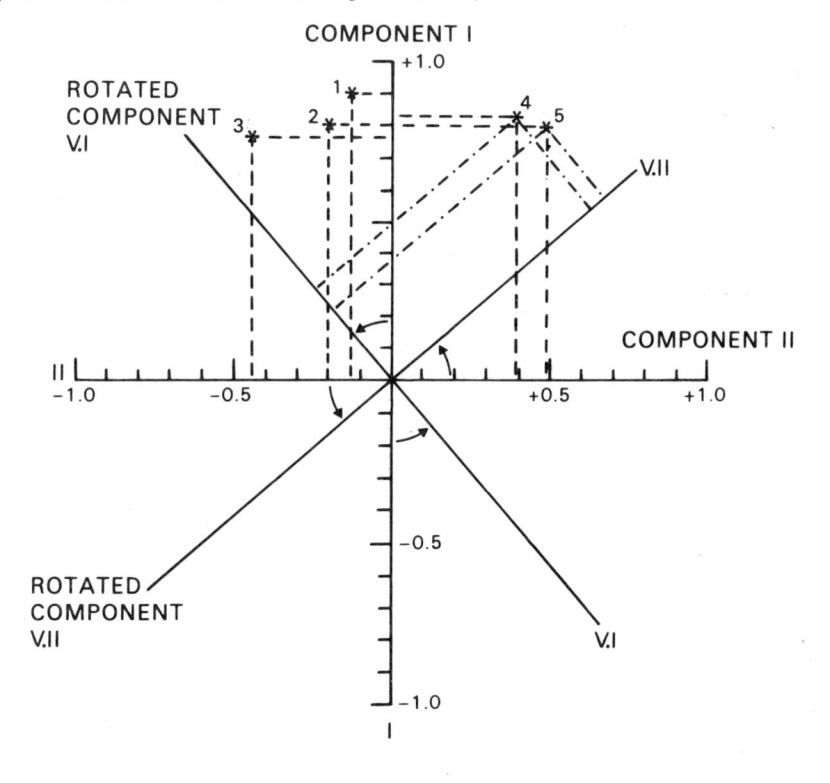

Fig. 9 Orthogonal rotation of overspill data. (dashed lines, loadings on principal components, dotted lines, loadings on Varimax rotated components, V.1 and V.II) (Source: Davies, 1971)

A. Unrotated component loadings

		Component			Pattern of high loadings		
		I	II	III	I	II	III
Aldridge Scheme	1	0.93	-0.16	-0.26	*		
	2	0.83	-0.23	-0.23	*		
	3	0.77	-0.45	-0.13	*	*	
Macclesfield Scheme	4	0.86	0.36	0.13	*		
	5	0.81	0.46	0.46	*	*	*

B. Varimax rotated component loadings

		Component			Pattern of high loadings		
		I	II	III	I	II	III
Aldridge Scheme	1	0.89	0.34	0.15	*		
	2	0.85	0.28	0.08	*		
	3	0.87	0.16	-0.15	*		
Macclesfield Scheme	4	0.36	0.91	0.04		*	
	5	0.21	0.99	-0.01		*	

Table 6: Unrotated and rotated component loadings: housing schemes in Aldridge and Macclesfield (Source Davies, 1971)

Re-examination of the original example of overspill schemes in the two towns will illustrate this point. It will be recalled that one component was extracted that was labelled "the degree of similarity" between the two schemes. A second component could be extracted on which schemes 1, 2 and 3 had low negative loadings and schemes 4 and 5 low positive loadings (Table 6). Such a component is referred to as bi-polar and is often difficult to interpret. However, if we rotate the original components about their origin two "new" components can be identified in which schemes 1, 2 and 3 load highly on factor 1 and schemes 4 and 5 on factor 2 (Figure 9 and Table 6). The rotation is performed without changing the position of the original variables; the co-ordinate axes are merely transformed. In this example the uncorrelated (or orthogonal) characteristics of the original components are also preserved.

PRINCIPAL COMPONENT SET

F_{1R} F_{2R}

ROTATED FACTORS SET 1 AND 2

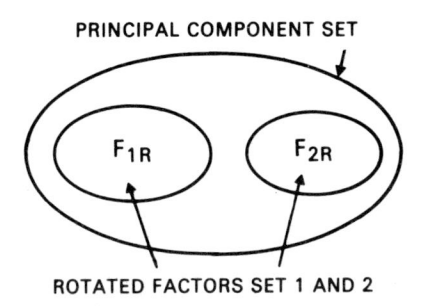

Fig. 10 Principal components and rotated factor subsets

In carrying out this rotation the general concept of the degree of similarity between the overspill scheme has been broken down into two more specific sub-concepts which in this case separate out the overspill schemes in the towns into two distinctive classes. Figure 10 attempts to summarise this point in the form of a Venn diagram.

Whether such a breakdown of the generality of a principal component is desirable depends entirely on the purpose of the investigation. In the study of overspill schemes the requirement was for a general index of similarity. In another study aimed at a functional regionalisation of Central London on the basis of data on inter-zonal taxi flows specificity was the objective (Goddard 1970). The principal component simply identified those zones that were linked by large volumes of traffic; analytic rotation divided these into two meaningful functional sub-regions. In so doing some of the variance associated with the principal component was redistributed between other components. This is illustrated in Table 7.

The usual criterion for rotation is that of simple structure, namely that each variable loads highly on one and only one factor, (Table 8). This can be achieved by the use of a number of algebraic (as opposed to graphical) criteria of which the most widely adopted is referred to as the "normal varimax criterion". As the name implies this seeks to maximise the variance of the loadings on each factor, that is to achieve as many high and as many low loadings as possible.

Using the varimax criteria the orthogonality of the original factors is maintained. However, in reality we might expect that the basic factors underlying the observed intercorrelation amongst our original variables may themselves be intercorrelated. For example, in the studies of the factorial ecology of a city it would be reasonable to assume that a factor describing the economic status of areas may be related to another describing their family status. This consideration led Bell to attempt to achieve a more meaningful description of urban spatial structure using an oblique rotation of an initial factor analysis. (Bell, 1955). The result of an application of an oblique rotation procedure to the analysis of social economic data for the Newcastle wards is described in Table 9 and Figure 11. From this analysis it will be seen that there is indeed a small correlation between the two factors originally identified ($R = -0.12$). It will also be seen that this rotation does bring

Component	I	II	III	IV	V	VI	Total
Eigenvalue	18.0	7.2	4.8	4.4	3.9	2.9	
% explanation	26.0	10.5	7.0	6.3	5.6	4.2	59.7
Rotated component	I	II	III	IV	V	VI	Total
Sum of squared loadings	10.8	8.3	6.9	6.0	5.4	3.8	
% explanation	15.7	12.0	9.9	8.7	7.9	5.6	59.7

Table 7: Explained variation associated with rotated and unrotated component analysis (Source: Goddard, 1970)

	Unrotated Factors				Simple Structure Factors		
Variables	I	II	III	Variables	I	II	III
1	*	*		1	*		
2	*	*	*	2	*		
3	*	*	*	3	*	*	
4	*		*	4	*		
5	*	*	*	5	*		
6	*	*		6	*		*
7	*		*	7			*

* Indicates a high factor loading.

Table 8: Hypothetical example of simple structure

us closer to the ideal of simple structure. Oblique factor 1 is more clearly identified as a social class factor and oblique factor 2 as a family status factor. This is principally because the rotation has succeeded in reducing the loadings of variable 4 (number of children under 5) and variable 5 (economically active women) on factor 1 and increasing them on factor 2, although the signs of the loadings of the two variables on factor 2 are obviously different.

Oblique rotation is conceptually attractive because in theory it can be shown to be the more general case; if the original clusters of variables are indeed orthogonal, an oblique rotation procedure should identify uncorrelated factors. Furthermore, with oblique rotation it is possible to carry out a higher order factor analysis of the correlations between the oblique factors themselves, thereby identifying more general underlying dimensions - although there is a danger here of reproducing one's original (general) principal factor.

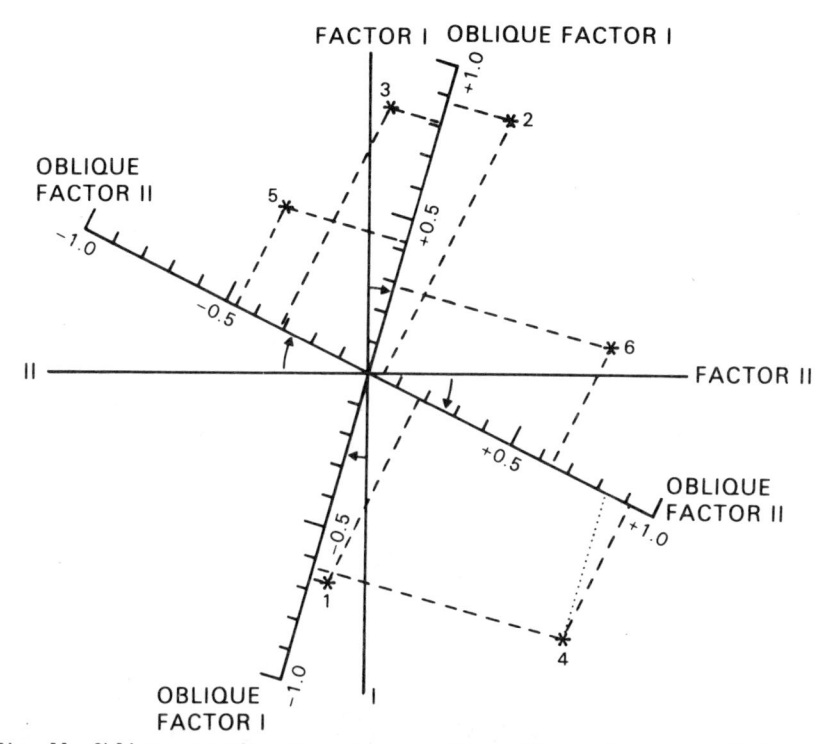

Fig. 11 Oblique rotation of social variables, City of Newcastle, 1971.
(dashed lines, <u>structure</u> loadings, dotted line, <u>pattern</u> loadings).

Against these advantages there are a number of analytical problems. Unlike orthogonal rotation, there can be no unique solution; while algebraic procedures are available the degree of obliqueness permitted is a matter for the investigator to specify. Also, once orthogonality is relaxed, the 1 to 1 correspondence between factor loadings and correlation coefficients is lost. Re-examination of figure 11 will reveal that in fact two sets of "loadings" can be identified. Those that are given in their entirety are the correlations between the variables and the factors since they are perpendicular projections of the variables onto the oblique factors. These are referred to as <u>structure loadings</u>. However, a second set of loadings, referred to as <u>pattern loadings</u>, could be derived; in the case of pattern loadings on oblique factor 1 these projections would lie parallel to oblique factor 2. For clarity, only one such pattern loading is shown on Figure 11.

A further complication is that a second set of underlying <u>reference axes</u> at right angles to the <u>primary</u> oblique factors can be constructed and a further set of pattern and structure loadings calculated. The primary reference axes are sometimes the most appropriate for delimiting simple structure.

29

Variable	Unrotated		Oblique rotated	
	I	II	I	II
1. Manual occupation	-0.65	-0.14	0.66	-0.13
2. A level education	0.79	0.42	-0.88	-0.06
3. Rateable value	0.82	0.09	-0.81	0.25
4. Young children	-0.81	0.57	0.60	-0.88
5. Econ. active women	0.53	-0.25	-0.43	0.46
6. Shared dwellings	0.11	0.76	-0.33	-0.67

$(r = -0.12)$

Note: No sigificance is attached to the reversal of signs of variables between the rotated and unrotated solutions if the relationship of variables one to another within a factor remains unaltered (i.e. -, +, +, -, +, + has the same interpretation as +, -, -, +, -, -).

Table 9   Unrotated and oblique rotation of social variables, City of Newcastle, 1971.

In view of the large number of rotational procedures and sets of loadings that can be adopted, considerable care needs to be taken when interpreting the results of an oblique factor analysis. All too often results are published without careful specification of the particular procedures adopted and comparisons of the results that would be obtained by the use of different procedures. In this confusing situation it is not surprising that many investigators prefer to stick to normal varimax rotation.

(vi)   Factor scores

The interpretation of R mode factor analysis in geography is often aided by examination of component or factor scores; that is the score of each observation on the various factors. Table 10 gives the raw data, Z scores and factor scores for two of the wards from the Newcastle example. Thus Dene ward (labelled C in Figure 2) has only 9.5% of its working population in semi and unskilled manual occupations, a relatively high proportion (6.9%) of its population possess one or more 'A' levels and the rateable value of property is 1.76 standard deviations above the average for all wards in the city. As a result this ward had a standard score of +1.98 on unrotated factor I (which we labelled social status). Conversely, Stephenson ward (labelled B in Figure 2) has 35.8% of its working population in semi or unskilled manual occupations, a low proportion possessing one or more 'A' levels (2.1%), the rateable value is -0.62 standard deviation below the average for all wards in the city, and nearly 10% of its households share a dwelling. Consequently this ward has a high negative score on factor 1 (-1.60). Scores of these wards on factor 2 are reversed but bear approximately the same relationship to each other as on factor 1. It would be reasonable to conclude that these two wards occupy fundamentally different positions in socio-economic space.

		Stephenson Ward (B)	Dene Ward (C)
Manual Occupations (per thousand)	Raw data	358	95
	Z	+1.07	-1.41
A level education (per thousand)	Raw data	21	69
	Z	-0.49	+0.89
Mean Rateable Value (£)	Raw data	111	189
	Z	-0.62	+1.76
Young children (per thousand)	Raw data	778	276
	Z	+2.0	-1.93
Econ. Active Women (per thousand)	Raw data	290	348
	Z	-1.39	+0.24
Shared Dwellings (per thousand)	Raw data	99	01
	Z	+2.89	-0.63
FACTOR SCORES			
Factor 1		-1.08	+1.98
Factor 2		+1.68	-1.06

Table 10 Raw data, Z scores and factor scores for selected wards, City of Newcastle, 1971.

Having said this, we should point out that considerable care must be taken in the evaluation of factor scores since they are derived on the basis of an observation's score on <u>all</u> variables not just those that may be highlighted in the interpretation of the factors (i.e. those with high factor loadings (See Horn 1973; Joshi 1972). In the case of components analysis the scores of a particular observation on a component are obtained by multiplying the observation's standardised scores on the original variables by the appropriate column of the component loading matrix.
The algebraic form is:

$$Sik = \sum_{J=1}^{m} Dij.Cjk$$

where Sik = component score for i th observation on k th component.
Dij = standard (z) score for i th observation on j th variable.
Cjk = component loading of j th variable on k th component.
m = number of original variables.

From this equation it should be clear that each score is the sum of a series of small products and that similar final sums can be arrived at in a variety of ways. Horn gives a simple example to illustrate this point:

$$\text{(i)} \quad Sik = (1, 0, 1, 0, 1) \begin{matrix} Dij \\ \end{matrix} \quad \begin{pmatrix} Cjk \\ 1 \\ 0 \\ 1 \\ 1 \\ 0 \end{pmatrix} = 2.0$$

$$\text{(ii)} \quad Sik = (0, 1, 1, 1, 0) \begin{matrix} Dij \\ \end{matrix} \quad \begin{pmatrix} Cjk \\ 1 \\ 0 \\ 1 \\ 1 \\ 0 \end{pmatrix} = 2.0$$

The implication is that observations with similar component scores will not necessarily have the same profile in terms of the raw variables. This point should be borne in mind when using component scores as the basis for sampling areas for more detailed investigation. The simple message, as with all statistical techniques, is that careful examination of the raw data should be a prelude to more advanced analysis.

A final warning before leaving this subject. In components analysis all of the variation in the original variables is accounted for by the components so we can therefore completely recreate our original observation in terms of our newly derived components simply by carrying out the simple matrix multiplication that we have described. However, in factor analysis, where there is some residual or unexplained variation, we have to <u>estimate</u> the scores of each observation on the factors. These estimates are obtained by the use of multiple regression techniques.

V SOME GEOGRAPHICAL EXAMPLES OF FACTOR ANALYSIS

In sections 1 and 2 we attempted to contrast two approaches to factor analysis, namely an inductive (descriptive) approach and a deductive (hypothesis-testing) approach. These are probably over-pompous terms for what may seem in the end to be two rather descriptive pieces of work. Nevertheless, the perspectives adopted were somewhat different. Rummel has suggested a tenfold classification of the types of research strategy within which factor analysis techniques may be employed (Rummel 1970). In this section we shall attempt to illustrate this classification by drawing upon examples from the geographical literature (Table 11). Our aim in doing this is merely to suggest the range of contexts in which factor analysis has been utilised in geography. We appreciate that other authors might have selected different examples or categorised the ones that we have taken differently. We hope that readers will follow up the examples for themselves if only to see whether they agree with our classification.

The first illustration considers <u>inter-relationships</u> within a data set, where there exist a number of covariant variables. The examples chosen are the overspill schemes discussed by Davies in section 1, where we were able to express the original correlation matrix in terms of two components, whilst losing only a small amount of the information within the data.

See Footnote page 36.

Table 11 A classification of factor analysis research strategies illustrated by geographical examples.

Strategy	Example	Subject	Data
Inter-relationship	Davies (1971)	Overspill schemes	59 questionnaire items, 5 overspill schemes.
Parsimony	Moser & Scott (1961)	Classification of British towns	60 census variables for 1951.
Structure	Sweetser (1969)	Comparison of ecological structure in Boston & Helsinki	Census data. 34 variables.
Classification	Davies (1972 (a))	Analyses of flow data to define regional & functional structure	Shopping trips. 108 settlements & 40 centres, S. Wales.
Scaling	Robson (1969)	Creation of taxonomic units within which to assess educational achievements, etc.	Census data, 30 variables for Sunderland.
Hypothesis-Testing	Bell (1955)	Creation of dimensions of social structure to justify hypothesis.	Seven census variables for L.A. 570 census units
Data-Transformation	Kirby & Taylor (1976)	Creation of orthogonal regression inputs in referendum analysis	Eight variables for 23 G.B. regions.
Exploratory	Goddard (1970)	Study of taxi flows within London to identify functional regions.	Taxi flows within London.
Mapping	Rees (1970)	Examination of social space in Chicago.	57 census variables.
Theory	Openshaw (1973 (B))	Rotation by criteria of predetermined spatial patterns of factor scores	Morphological data, S. Shields.

The second category relates to parsimony, or the reduction of a large number or variables to manageable proportions. One of the best examples is provided by Moser & Scott (1961), who classified 157 British towns in relation to 60 census variables. Naturally, a classification based upon four factors, rather than 60 variables, is both speedier and more meaningful.

At this point we should emphasise that factor analysis will not produce a classification of our observations in the strict sense of assigning them to discrete classes. For this we need to revert to some grouping algorithm. One of the reasons Moser & Scott sought to reduce their number of variables was that they had no such algorithm available and were forced to subjectively group British towns on the basis of common patterns of scores on the factors. Despite this, Moser & Scott's study remains as one of pioneer uses of factor analysis in the social sciences.

Rummel defines structure in terms of analysis of the common elements within different samples. An interesting example of this is provided by Sweetser (1969), who performed a factorial ecology on data for both Boston and Helsinki. Whilst socio-economic status and family status were of similar importance in both cities, Sweetser noted that unique dimensions of career women (Helsinki) and ethnicity (Boston) existed.

The fourth category relates to the widest use of factor methods, notably the creation of weighted scales by which to score the base units, in order to create like regions. This strategy has been employed numerous times in the examination of factorial ecologies of many cities, (for a full bibliography see Robson (1973)). One of the earliest and most detailed studies was undertaken by Robson (1969). Here, a principal components analysis was undertaken upon 30 census variables, each measuring some aspect of economic structure or housing type for each of 263 enumeration districts in Sunderland. Four dimensions resulted, which measured social class, housing quality, tenure and a generalised poverty factor. Each component was multiplied by the original data to give component scores, which were then combined into homogenous groups using cluster analysis. These in turn became the new base units for a study of attitudes to education within the city.

Davies provides another example of factor analysis, with his study of shopping trips (Davies 1972(a)), which we use to illustrate a classificatory strategy. Here, a Q-mode analysis (using places for variables within the initial data matrix) was undertaken upon the flows between 108 settlements and 40 shopping centres in South Wales. The resulting ten components picked out the settlements with similar flow patterns, whilst the scores showed the patterns of spatial association between centres and settlements. These patterns of shopping trips were then used to define functional regions for further analysis. It is to be noted that although both Robson and Davies used factor methods to create taxonomic units, their methods were rather different. Robson's research involved two stages: the creation of scales (the components) which measured aspects of social differentiation, and the mapping of the scores of areas on these scales. Davies, by way of contrast, had a finite data set of observations, and the problems of choosing meaningful census variables, or interpreting (or rejecting) factors did not arise.

We take as our example of hypothesis-testing the work of Bell (1955) that we have already considered in section (ii). It remains one of the clearest examples of quantitative analysis being used to test an *a priori* (or previously developed) hypothesis.

The seventh category derived by Rummel is that of data transformation. A simple example is provided by Kirby & Taylor (1976). Here, the authors attempted to explain the variation of the 1975 E.E.C. Referendum vote in 23 regions of the U.K., in terms of one (or both) of the following hypotheses: namely that the vote was a result of normal political cleavages, or a locational factor, whereby distance from government produced an alienation effect. Eight variables were chosen to represent shades of these hypotheses, and a factor analysis undertaken to produce two (uncorrelated) varimax factors. The rotation was necessary to satisfy the requirements of the linear regression model, by which the voting data was compared with the locational factor, and the political factor scores. The regression results indicated that the locational factor could be removed from the analysis.

The category of exploratory use relates to areas in which little work has been undertaken, and where the researcher has no opportunity to undertake controlled laboratory experiments, a situation which is normally the case in social science. The example chosen here is Goddard's 1970 study of taxi flows within London, which was a preliminary to investigating the functional structure of a metropolitan centre. Once more, the analysis is similar to that undertaken by Davies, in so far as classification is a final step.

The term mapping may cause some confusion in the geographic context. Rummel's term relates to the reworking of known empirical concepts in different situations in an attempt to discover sources of variation and common elements. Again, factorial ecologies are the widest example of this approach, whilst Rees' study of Chicago is perhaps the most detailed published examination of the social space of a major city.

The concluding section concerns the use of factor methods in a theoretical context, i.e. utilising the linear algebra within the model to make rigorous mathematised deductions about the data under analysis. Such high levels of research are rare in geographical analysis, and one of the very few examples is provided by Openshaw (1973(B)). Openshaw has been concerned with the effects of scale on relationships, a fundamental concept elaborated by Blalock (1964), and the effects of autocorrelation within data, an equally fundamental consideration, discussed fully by Cliff & Ord (1973). These interests have led him to argue that it is possible to rotate factors to situations in which the spatial autocorrelation of factor scores is at an extreme, thereby providing information concerning the relationships between the variables, and their spatial patterns. He has also performed factor analyses using data from different size lattices, and examined the resulting variations in loadings and eigenvalues (Openshaw 1973(a)).

It is to be noted that these examples have several similar elements, and indeed, Rummel's classification could easily be compressed, or extended. Nevertheless, the sheer scope of the technique underlines the importance of our interest in factor methods.

VI CONCLUDING REMARKS

This monograph can only serve as an introduction to factor analysis. We are conscious of the fact that we have only discussed two members of the factor analysis family, namely principal components analysis and common factor analysis, and that there is a rapidly widening range of solutions, such as image analysis and alpha analysis, many of which bypass some of the methodological difficulties that we have touched upon. We have also avoided discussing in any detail the various analytical rotation procedures that go under such glorious names as "oblimin" and "biquartimin". But these are all issues to be taken up in a more advanced monograph. All we have attempted is to describe "old-fashioned" factor analysis as it has been used and still continues to be used in geography.

"Fashion" is a good note to conclude on. Factor analysis has been very much in fashion amongst geographers, but many of those who leaped onto the bandwagon in the early days have now jumped off. Those who remain see factor analysis as a useful tool in certain contexts, to be taken up after only carefully exhausting what can be gained from the data by simpler forms of analysis. Perhaps one of the reasons why factor analysis is likely to remain an attractive tool for some geographers is the fact that it seldom produces "right" answers; a great deal of intuition and a large number of computer runs are usually needed to produce a "satisfactory" result (if we are honest, most would agree that research in fact progresses in this way).

Our final stricture will either make the reader a factor analyst for life or cause him or her to abandon the technique for ever. This is to grab some census data for your town, a factor analysis program and a large computer and run the data through the program ten, twenty, maybe thirty times, trying all of the options. But be warned! Most computer centres are rationing computer paper these days.

FOOTNOTE

The regression equation for estimating a matrix of factor scores from the data in matrix form is:

$$\hat{S}\ (n \times p) = Z(n \times m) . \beta\ (m \times p)$$

where: $\hat{S}\ (n \times p)$ = regression estimates of n common factor scores on each of p common factors

$\beta\ (m \times p)$ = p columns of m regression coefficients, where ℓ th column gives the m regression coefficients for estimating the n factor scores on the ℓ th factor.

$Z\ (n \times m)$ = standardized data matrix of n observations and the variables.

REFERENCES CITED WITHIN THE TEXT

Bell, W. (1955), Economic, family and ethnic status: an empirical test. *American Sociological Review*, 20, pp 45-52.

Berry, B.J.L. (1960), An inductive approach to the regionalization of economic development and cultural change. in: *Essays on geography & economic development*, (ed) N. Ginsberg, (Research Paper 62, Department of Geography, University of Chicago).

Berry, B.J.L. & Horton, F.E. (1970), *Geographical perspectives on urban systems*. Englewood Cliffs.

Blalock, H.M., Jr. (1964), *Causal inferences in non-experimental research*. University of North Carolina Press, Chapel Hill.

Britton, J.N.M. (1968), *Regional analysis & economic geography - A case study of the Bristol Region*. G. Bell & Sons, London.

Child, D. (1970), *The essentials of factor analysis*. Holt, Rinehart & Winston.

Clark, D. (1975), Understanding canonical correlation analysis. *CATMOG*, 3, (Geo-Abstracts Ltd)

Cliff, A.D. & Ord, K. (1973), *Spatial autocorrelation*. Pion Press.

Davies, W.K.D. (1971), Varimax and the destruction of generality - a methodological note. *Area*, 3, pp 112-8.

Davies, W.K.D. (1972), Conurbation and city region in an administrative borderland; a case study of the greater Swansea area. *Regional Studies*, 6, pp 217-36.

Goddard, J.B. (1968), Multivariate analysis of office location patterns in the city centre: a London example. *Regional Studies*, 2, pp 69-85.

_____ (1970), Functional regions within the city centre: a study by factor analysis of taxi flows in central London. *Transactions Institute of British Geographers*, 49, pp 161-182.

Gould, P.R. (1967), On the geographic interpretation of eigenvalues. *Transactions Institute of British Geographers*, 42, pp 53-86.

Gould, P.R. & White, R.R. (1968), The mental maps of British school leavers. *Regional Studies*, 2, pp 161-182.

Gregory, S. (1963), *Statistical methods & the geographer*. Longmans, London.

Harman, H.H. (1960), *Modern factor analysis*. University of Chicago Press.

Hay, A. (1975), On the choice of methods in the factor analysis of connectivity matrices - a comment. *Transactions Institute of British Geographers*, 66, pp 163-167.

Horn, C.J. (1973), Factor scores and geographical research. *Institute of British Geographers, Quantitative Methods Study Group, Working Paper I*, pp 26-29.

Horst, P. (1965), *Factor analysis of data matrices*. Holt, Rinehart & Winston.

Joshi, T.R. (1972), Towards computing factor scores.in: *International geography*, (ed) W.P. Adams & F.M. Helleiner, (Montreal I.G.U.)

King, L.J. (1969), Cross-sectional analysis of Canadian urban dimensions, 1951 and 1961. *Canadian Geographer*, 10, pp 205-224.

Kirby, A.M. (1976), A three-dimensional model for the introductory teaching of multivariate analysis. *International Journal of Mathematical Education in Science & Technology*. (forthcoming)

Kirby, A.M. & Taylor, P.J. (1976), A geographical analysis of the voting pattern in the E.E.C. referendum June 5, 1975. *Regional Studies*, (forthcoming).

Knox, P.L. (1974), Spatial variations in levels of living in England and Wales in 1961. *Transactions Institute of British Geographers* 62, pp 1-24.

Moser, C.A. & Scott, W. (1961), *British towns: a statistical study of their social and economic differences*. Oliver & Boyd.

Openshaw, S. (1973 a), An empirical study of the influence of scale on principal component and factorial studies of spatial association. Paper presented to Institute of British Geographers, January, 1973.

Openshaw, S. (1973 b), Some ideas concerning a spatial criterion for geographical factor analysis. Paper presented to Institute of British Geographers, August, 1973.

Rees, P.H. (1970), The factorial ecology of Chicago. in: *op. cit*, (ed) B.J.L. Berry & F.E. Horton.

Robson, B.T. (1969), *Urban analysis: a study of city structure with special reference to Sunderland*. Cambridge University Press.

———— (1973), A view on the urban scene. in: *Studies in human geography*, (eds) M.K. Chisholm & B. Rodgers, Heinemann.

Rummel, R.J. (1970), *Applied factor analysis*. Northwestern University Press.

Russett, B.M. (1967), *International regions and the international system: a study in political ecology*. Rand McNally.

Sweetser, F. (1969), Ecological factors in metropolitan zones and sectors. in: *Quantitative ecological analysis in the social sciences*, (eds) M. Dogan & S. Rokkan, M.I.T.

Taylor, P.J. (1975), *Distance decay in spatial interactions CATMOG* 2. (Geo Abstracts Ltd)

Tinkler, K.J. (1972), The physical interpretation of eigenfunctions of dichotomous matrices. *Transactions Institute of British Geographers*, 55, pp 17-46.

———— (1975), On the choice of methods in the factor analysis of connectivity matrices - a reply. *Transactions Institute of British Geographers*, 66, pp 168-170.

REVIEWS

Clark, D., Davies, W.K.D. & Johnston, R.J. (1974), The application of factor analysis in human geography. *The Statistician*, 23(3/4), pp 259-281.

Janson C-G. (1969), Some problems of ecological factor analysis: in: *op cit.* (ed) M. Dogan & S. Rokkan.

Krumbein, W.C. & Graybill, F.A. (1965), *Statistical models in geology*. McGraw-Hill.

Mather, P.M. & Doornkamp, J.C. (1970), Multivariate analysis in geography with particular reference to drainage basin morphometry. *Transactions Institute of British Geographers*, 51, pp 163-187.

Rees, P.H. (1971), Factorial ecology: an extended definition, survey and critique of the field. *Economic Geography*, 47, pp 220-233, (supplement).